# YOUR KNOWLEDGE HAS VALUE

AF131319

Bradley Tice

# A Radix 4 Based System for Use in Theoretical Genetics

GRIN Verlag

**Bibliografische Information der Deutschen Nationalbibliothek:**

Die Deutsche Bibliothek verzeichnet diese Publikation in der Deutschen National-
bibliografie; detaillierte bibliografische Daten sind im Internet über http://dnb.d-
nb.de/ abrufbar.

**Imprint:**

Copyright © 2008 GRIN Verlag GmbH
Druck und Bindung: Books on Demand GmbH, Norderstedt Germany
ISBN: 978-3-656-64524-5

**This book at GRIN:**

http://www.grin.com/en/e-book/198601/a-radix-4-based-system-for-use-in-theore-
tical-genetics

# A Radix 4 Based System for Use in Theoretical Genetics

By Bradley S. Tice
Advanced Human Design, P.O. Box 3868, Turlock, California 95381 U.S.A.

## Abstract

The paper will introduce the quaternary, or radix 4, based system for use as a fundamental standard beyond the traditional binary, or radix 2, based system in use today. A greater level of compression is noted in the radix 4 based system when compared to the radix 2 base as applied to a model of information theory. The application of this compression algorithm to both DNA and RNA sequences for compression will be reviewed in this paper.

Keywords: Radix 4, Quaternary, Theoretical Genetics, DNA Compression, RNA Compression

## I. Introduction

A quaternary, or radix 4 based system, is defined as four separate characters, or symbols, that have no semantic meaning apart from not representing the other characters. This is the same notion Shannon gave to the binary based system upon it's publication in 1948 [1]. This paper will present research that shows the radix 4 based system to have a compression value greater than the traditional radix 2 based system in use today [2]. The compression algorithm will be used to compress DNA and RNA sequences. The work has applications in theoretical genetics and synthetic biology.

## 2. Randomness

The earliest definition for randomness in a string of 1's and 0's was defined by von Mises, but it was Martin-Lof's paper of 1966 that gave a measure to randomness by the *patternlessness* of a sequence of 1's and 0's in a string that could be used to define a random binary sequence in a string [3 and 4]. A non-random string will be able to compress, were as a random string of characters will not be able to compress. This is the classical measure for Kolmogorov complexity, also known as Algorithmic Information Theory, of the randomness of a sequence found in a binary string.

## 3. Compression Program

The compression program to be used has been termed the *Modified Symbolic Space Multiplier Program* as it simply notes the first character in a line of characters in a binary sequence of a string and subgroups them into common or like groups of similar characters, all 1's grouped with 1's and all 0's grouped with 0's, in that string and is assigned a single character notation that represents the number found in that sub-group, so that it can be reduced, compressed, and decompressed, expanded, back to it's original length and form [5]. An underlined 1 or 0 is usually used to note the notation symbol for the placement and character type in previous applications of this program. The underlined initial character to be compressed will be used for this paper.

## 4. Application of Theory

The application of a quaternary, or radix 4 based system, to existing genetic marking and counting systems has many advantages. The first is the greater amount of compression from this base, as opposed to the standard binary based system in use today, and secondly, as a more utilizable system because of the four character, or symbol, based system that provides for more variety to develop information applications.

# 5. DNA

DNA, or Deoxyribonucleic acid, is a linear polymer made up of specific repeating segments of phosphodiester bonds and is a carrier of genetic information [6]. There are four bases in DNA; adenine, thymine, guanine and cytosine [7].

The use of a compression algorithm for sequences of DNA.

Definitions
A = Adenine
T = Thymine
G = Guanine
C = Cytosine

Example #A

ATATGCGCTATACGCGTATATATA

The compression algorithm will use a specific focus on TA and GC DNA sequences in Example #A.

Key Code

TA = 4 characters

GC = 2 characters

Compress Example #A

ATAT<u>GC</u>ATATCGCG<u>TA</u>

The compressed DNA sequence is 16 characters from the original non-compression total of 24.

The use of a four character system, a radix 4 base number system, that is composed of each character not representing the other characters is ideal in DNA sequences composed of adenine, thymine, guanine and cytosine.

Example #D

TAGCTAGCTAGCTAGCTAGCTAGCTAGCTAGCTAGCTAGC

Key Code

TAGC = 10

Compression of Example #D

TAGC

The compressed version of Example #D is 4 characters from the original non-compressed total of 40 characters.

# 6. RNA

RNA, or Ribonucleic acid, translates the genetic information found in DNA into proteins [8]. There are four bases that attach to each ribos [9].

2

The use of a compression algorithm for sequences of RNA.

Definitions
A = Adenine
C = Cytosine
G = Guanine
U = Uracil

Example #B

AUAUCGCGAUAUCGCGUAUAUAUAGCGC

The compression algorithm will focus on specific RNA sequences.

Key Code

UA = 4 characters

GC = 2 characters

Compress Example #B

AUAUCGCGAUAUCGCG<u>UAGC</u>

The compressed RNA sequence is 20 characters in length form the original non-compression total character length of 28.

The use of a four character system, a radix 4 base number system,   that is composed of each character not representing the other characters is ideal in RNA sequences composed of adenine, cytosine, guanine and uracil.

Example #C

UAGCUAGCUAGCUAGCUAGCUAGC

The use of a universal compression algorithm is as follows:

Key Code

UAGC = 6

Compression of Example #C

<u>UAGC</u>

The compressed version of Example #C is 4 characters from the original non-compressed 24 character total length.

## Summary

The compression algorithm used for both DNA and RNA sequences has the power of both a universal compression algorithm, all character length types, and a specific, or target, level of compression.

## References

1. "Shannon, C.E., "A Mathematical Theory of Information", *Bell Labs. Tech. Jour*. 27, 379-423 and 623-656 (1948).

2. Tice, B.S., "The analysis of binary, ternary and quaternary based systems for communications theory", Poster for the SPIE Symposium on Optical Engineering and Application Conference, San Diego, California, August 10-14, 2008.

3. Kotz, S. and Johnson, N.I., Encyclopedia of Statistical Sciences, John Wiley & Sons, New York (1982).

4. Martin-Lof, P., "The definition of random sequences", *Information and Control*, 9, pp. 602-619 (1966).

5. Tice, abide.

6. L.C. Lutter, "Deoxyribonucleic acid". In *McGraw-Hill Encyclopedia of science & technology*. McGraw-Hill Publishers, New York, pp. 373-379 (2007).

7. Lutter, abide., p. 374.

8. A.L. Beyer and M.W. Gray, "Ribosomes". In *McGraw-Hill Encyclopedia of science & technology*. McGraw-Hill Publishers, New York, pp. 542-546 (2007).

9. Beyer, abide., p. 542.